# Rico Cosby

## The Heroic Work
## of a
## Mine Detection Dog

Anne Wooleyhand

ISBN 978-1-62806-383-7  (print | paperback)

Library of Congress Control Number 2023911289

Published by Salt Water Media
29 Broad Street, Suite 104
Berlin, MD 21811
www.saltwatermedia.com

Salt Water
MEDIA

Cover image by the author

Interior images by the author and stock photos

# THE MARSHALL
# LEGACY INSTITUTE
Removing Landmines, Promoting Stability

The Marshall Legacy Institute (MLI) is a U.S. non-profit organization dedicated to helping countries affected by landmines. We provide highly trained dogs, including MDD Rico-Cosby, to countries around the world to help sniff out mines and save lives. The CHAMPS program gives kids a chance to become involved in this important issue, and many kids have sponsored dogs, like Rico, and have helped many landmine survivors get the assistance they need.

This book is dedicated to all the human handlers and their canine partners who place their lives in danger to locate and remove landmines making the world a safer place. These heroic men, women, and dogs have saved countless lives, and helped communities to be able to farm, walk, and play on land that was once contaminated by landmines. We are grateful to these heroes for their impressive service.

My name is Rico Cosby, and I am an eight-year-old Belgian Malinois.

I was born in the Netherlands on August 21, 2014.

The story of my life is interesting, and some people call me a hero.

In 2015, a young man named Henry Harris and students from The Brunswick School in Connecticut raised $20,000 to fund my training, food, medical care, toys, and transportation. Henry rode his bike across the country to help his school raise the money. They named me Cosby in memory of a beloved teacher, Mr. Cosby, at the Brunswick School who had passed away.

Dogs like me are very special.

When I was just a little puppy in Europe, I was selected to begin training so I could become a mine detection dog. I lived with my trainer for about a year and learned obedience, fetching, and retrieving skills.

As a one-year-old pup, I moved to the Mine Detection Dog Center in Bosnia-Herzegovina. This is where the trainers observed how well I followed directions, if I liked to sniff and play with my training toy—a ball or special chew toy—and how I behaved around other dogs and humans.

The veterinarians checked to be sure my hips and elbows were good and made sure I was in great health.

I also learned how to sniff different explosive odors and was rewarded with my special toy when I did a good job.

Other parts of my training included learning how to sniff in a straight line and how to sit, stay, and look when I detected the explosive odor.

This required a great deal of practice and hard work.

For the final part of my training, I was introduced to Kenan, my human handler.

He named me Rico. That is how I became Rico Cosby.

Kenan and I trained together and built a great friendship.

Once my training was completed, I had to take a test to become an accredited Mine Detection Dog.

My motivation and performance were evaluated daily to be sure I was ready to work sniffing out landmines.

I loved working because I would get to play with my special toy.

Kenan and I kept each other safe and worked together to save many lives in Bosnia-Herzegovina for six years.

You may wonder what landmines are and why they are a problem. Landmines are devices planted in or placed right on top of the ground that are designed to explode. They come in all different shapes and sizes. Some are small like a hamburger (anti-personnel landmines) while others are big like a tire (anti-tank landmines). Landmines contain dangerous explosives and can be made out of metal, wood, or plastic. They are used to keep enemies out of areas such as borders and to restrict the movement of opposing forces during conflict or war.

When a person or animal steps on a landmine, the pressure from their weight activates the mine. The problem is that when a conflict or war is over, mines are sometimes left in the ground and can remain a hidden threat for more than 50 years. Innocent people and animals step on the mines and are injured or killed, because they did not know the mines were there. The land is not safe to use for farming, recreation, or building. Children cannot run and play or walk to school without the danger of stepping on a landmine.

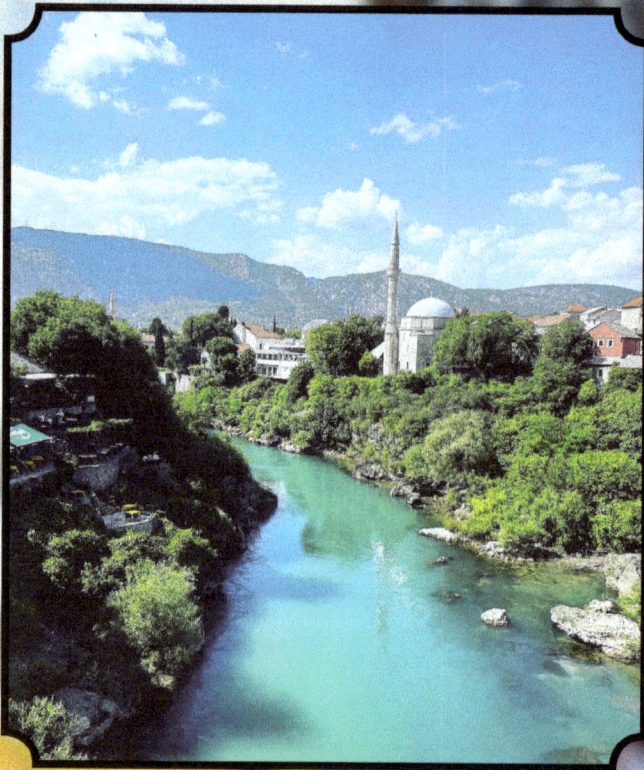

In my career working with Kenan, we assisted in the clearance of 600,000 square meters of land. By sniffing out landmines in Bosnia-Herzegovina, I saved lives and helped to make it safer for people to use the land.

Bosnia-Herzegovina has more landmines than any other country in Europe. It is a small but beautiful country with a rich culture. I loved living and working there.

Why are dogs like me used to sniff out landmines?

Well, I have a powerful sense of smell and can detect vapors released by landmines.

As part of my training, I learned to detect three specific chemicals found in landmines.

I worked much faster than a human probing the ground, and I detected landmines made from different materials such as plastic and metal.

Metal detectors are a tool often used in locating mines, but they cannot locate the mines made of plastic.

Safety is very important, and out of the 280 mine detection dogs MLI has provided, no dogs have been killed or injured while working in the field.

People often ask if sniffing out landmines is dangerous, and the answer is "yes". Each morning, I showed I was ready to work by searching an area of land for a training device that contained small amounts of explosive odor. I could not work that day unless I found the training device on my first try. Kenan and I would spend the day training if I did not pass the test in the field.

If it was too windy or if there was too much moisture in the air or on the ground, I did not work. These were strict protocols in place that kept Kenan and me safe. When I worked, I had to be focused and able to sniff out any landmines in the ground.

After six years of sniffing out landmines in Bosnia-Herzegovina, it was time for me to retire. Many mine detection dogs retire after six or eight years. Some have worked as long as ten or twelve years. With Kenan at my side, my retirement was officially announced and celebrated during a special ceremony. I met my new friends, Anne and Indre, right before the ceremony and learned that they were going to take me to my new home in the United States where I would become an Ambassador for Mine Detection Dogs everywhere. Anne is going to be my new handler in retirement, and I will live with her family.

The ceremony was very nice.

Following the ceremony, Kenan took me outside to show my new friends some fun tricks and how I love my ball. Kenan also met his new dog who will continue the work I was doing to actively sniff out landmines in Bosnia-Herzegovina.

I am so proud of the work Kenan and I did, and I know he will take great care of his new dog.

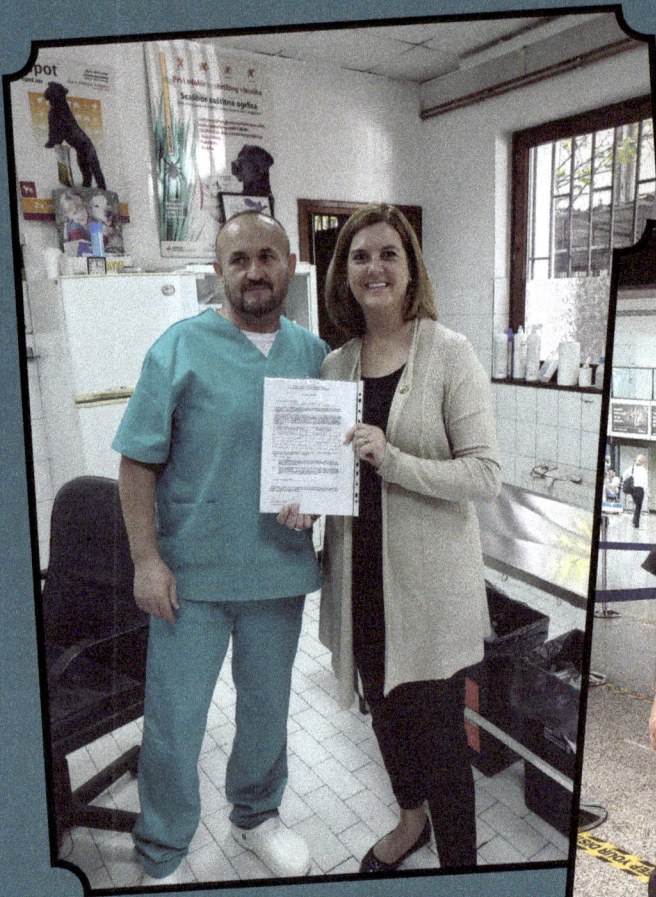

My time in Bosnia-Herzegovina came to an end, and it was time to go to the United States to become the eighth K9 Ambassador for the Marshall Legacy Institute.

Anne met with the veterinarian after the ceremony to sign my medical records and get my passport. Kenan and I met Anne and Indre at the airport a couple of days later so I could fly to my new home.

Saying goodbye to Kenan made me sad, and I will miss him very much.

Welcome to the United States, Rico!

It is time to start your new life as the *Marshall Legacy Institute's K9 Ambassador.*

# THE MARSHALL LEGACY INSTITUTE

Removing Landmines, Promoting Stability

# CHAMPS KIDS

CHAMPS (Children Against Mines Program) is a program offered by The Marshall Legacy Institute (MLI), a nonprofit organization committed to addressing the humanitarian impact of landmines.

CHAMPS is one of MLI's educational programs designed to help young people learn about landmines, the negative effect they have on communities, and how young people can take action towards having a positive influence in the world.

*Proceeds from the sale of this book will be used to sponsor a life-saving mine detection dog.*

Canine Ambassador

MDD RICO COSBY

www.ingramcontent.com/pod-product-compliance
Lightning Source LLC
Chambersburg PA
CBHW060946100426
42813CB00016B/2873